and Norman

by Mona Lee
illustrated by Barbara Hranilovich

Harcourt

Orlando   Boston   Dallas   Chicago   San Diego

www.harcourtschool.com

Oscar is a clown.
Norman wants to be a clown.

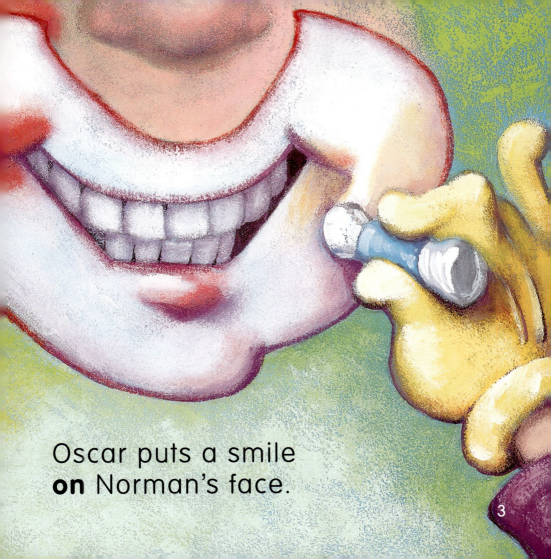

Oscar puts a smile **on** Norman's face.

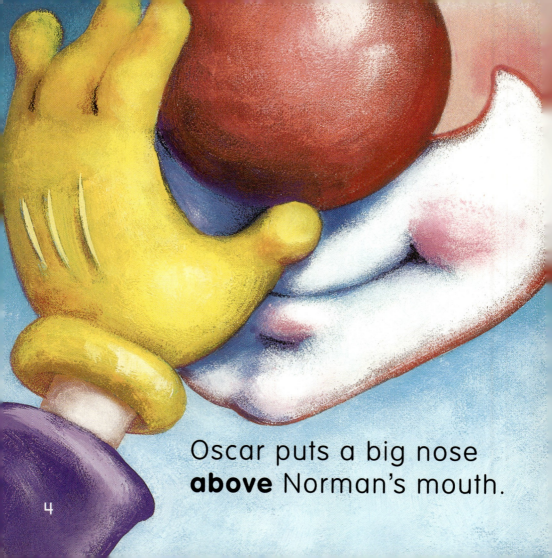

Oscar puts a big nose **above** Norman's mouth.

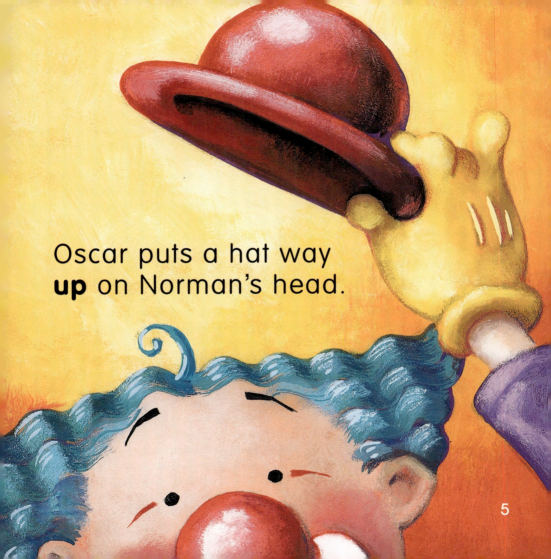
Oscar puts a hat way **up** on Norman's head.

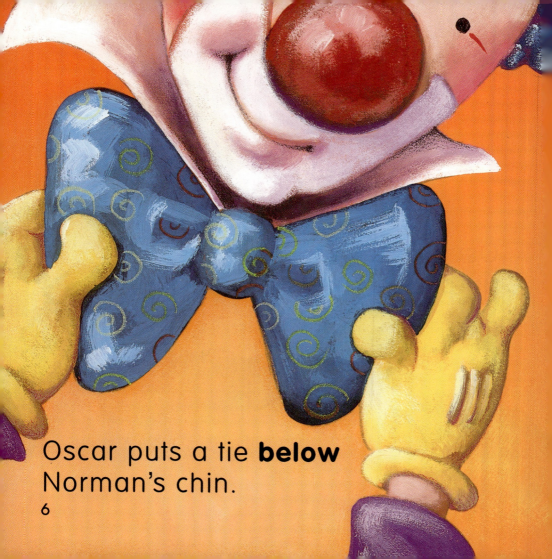

Oscar puts a tie **below** Norman's chin.

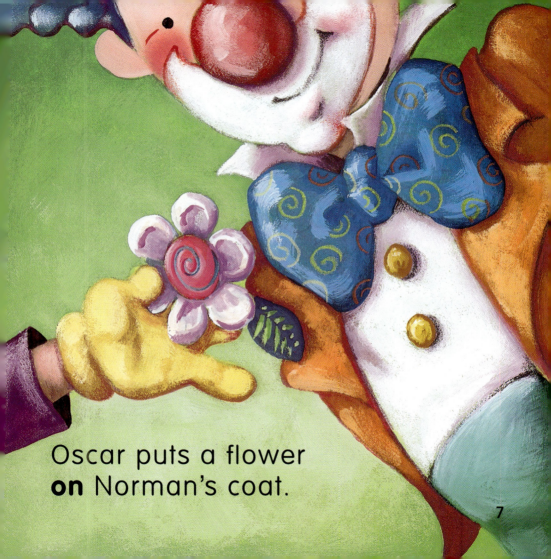

Oscar puts a flower **on** Norman's coat.

Oscar puts a scarf **inside** Norman's back pocket.

Oscar puts a glove on Norman's **left** hand.

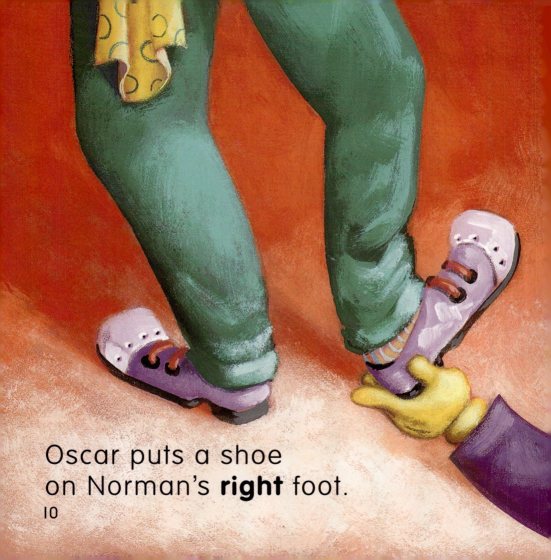

Oscar puts a shoe on Norman's **right** foot.

Oscar walks **next** to Norman.

It's showtime!